Developing Metrics for Assessing Engineering Instruction

What Gets Measured Is What Gets Improved

Report from the Steering Committee for Evaluating Instructional Scholarship in Engineering

NATIONAL ACADEMY OF ENGINEERING
OF THE NATIONAL ACADEMIES

THE NATIONAL ACADEMIES PRESS
Washington, D.C.
www.nap.edu

THE NATIONAL ACADEMIES PRESS 500 Fifth Street, N.W. Washington, DC 20001

NOTICE: The project that is the subject of this report was approved by the Governing Board of the National Academies, whose members are drawn from the councils of the National Academy of Sciences, the National Academy of Engineering, and the Institute of Medicine. The members of the steering committee responsible for the report were chosen for their special competencies and with regard for appropriate balance.

This is a report of work supported by the National Science Foundation through Grant No. 0633774. Any opinions, findings, conclusions, or recommendations expressed in this publication are those of the authors and do not necessary reflect the view of the organizations or agencies that provided support for the project.

International Standard Book Number 13: 978-0-309-13782-9

International Standard Book Number 10: 0-309-13782-9

Cover: Socrates and His Students, by Johann Friedrich Greutner, 17th century.

Copies of this report are available from the National Academies Press, 500 Fifth Street, N.W., Lockbox 285, Washington, DC 20055; (888) 624-8373 or (202) 334-3313 (in the Washington metropolitan area); Internet, http://www.nap.edu.

Printed in the United States of America

THE NATIONAL ACADEMIES

Advisers to the Nation on Science, Engineering, and Medicine

The **National Academy of Sciences** is a private, nonprofit, self-perpetuating society of distinguished scholars engaged in scientific and engineering research, dedicated to the furtherance of science and technology and to their use for the general welfare. Upon the authority of the charter granted to it by the Congress in 1863, the Academy has a mandate that requires it to advise the federal government on scientific and technical matters. Dr. Ralph J. Cicerone is president of the National Academy of Sciences.

The **National Academy of Engineering** was established in 1964, under the charter of the National Academy of Sciences, as a parallel organization of outstanding engineers. It is autonomous in its administration and in the selection of its members, sharing with the National Academy of Sciences the responsibility for advising the federal government. The National Academy of Engineering also sponsors engineering programs aimed at meeting national needs, encourages education and research, and recognizes the superior achievements of engineers. Dr. Charles M. Vest is president of the National Academy of Engineering.

The **Institute of Medicine** was established in 1970 by the National Academy of Sciences to secure the services of eminent members of appropriate professions in the examination of policy matters pertaining to the health of the public. The Institute acts under the responsibility given to the National Academy of Sciences by its congressional charter to be an adviser to the federal government and, upon its own initiative, to identify issues of medical care, research, and education. Dr. Harvey V. Fineberg is president of the Institute of Medicine.

The **National Research Council** was organized by the National Academy of Sciences in 1916 to associate the broad community of science and technology with the Academy's purposes of furthering knowledge and advising the federal government. Functioning in accordance with general policies determined by the Academy, the Council has become the principal operating agency of both the National Academy of Sciences and the National Academy of Engineering in providing services to the government, the public, and the scientific and engineering communities. The Council is administered jointly by both Academies and the Institute of Medicine. Dr. Ralph J. Cicerone and Dr. Charles M. Vest are chair and vice chair, respectively, of the National Research Council.

www.national-academies.org

Steering Committee for Evaluating Instructional Scholarship in Engineering

C. JUDSON KING, *chair*, University of California Berkeley

SUSAN A. AMBROSE, Carnegie Mellon University

RAOUL A. ARREOLA, University of Tennessee, Health Science Center

KARAN WATSON, Texas A&M University

Staff

RICHARD M. TABER, Study Director

NORMAN L. FORTENBERRY, Director, Center for the Advancement of Scholarship on Engineering Education

JASON WILLIAMS, Senior Program Assistant

ELIZABETH T. CADY, Associate Program Officer

TYLISHA BABER, Christine Mirzayan Science and Technology Policy Fellow

CAROL R. ARENBERG, Senior Editor

Acknowledgements

The committee wishes to thank the individuals that participated in the November 2007 workshop on instructional metrics. The conversations and insights gained in that workshop provided excellent guidance for the structure of this report. Special thanks are extended to the contributors of white papers for the workshop: Michael Theall (Youngstown State University), Lawrence M. Aleamoni (University of Arizona), and; Larry A. Braskamp (Loyola University of Chicago).

This report has been reviewed by individuals chosen for their diverse perspectives and expertise, in accordance with procedures approved by the National Academy of Engineering (NAE). The purpose of this independent review is to provide candid and critical comments that will assist the authors and the NAE in making the published report as sound as possible and to ensure that the report meets institutional standards for objectivity, evidence, and responsiveness to the study charge. The review comments and draft manuscript remain confidential to protect the integrity of the deliberative process.

We wish to thank the following individuals for their participation in the review of this report:

CRISTINA AMON, University of Toronto

MAURA BORREGO, Virginia Polytechnic Institute and State University

JOHN CENTRA, Syracuse University

ALAN CRAMB, Illinois Institute of Technology

THOMAS LITZINGER, The Pennsylvania State University

JACK LOHMANN, Georgia Institute of Technology
RICHARD K. MILLER, Franklin W. Olin College of Engineering

MICHAEL THEALL, Youngstown State University

REPORT REVIEW MONITOR

LOUIS J. LANZEROTTI, New Jersey Institute of Technology

Contents

Executive Summary

Faculty in all disciplines must continually prioritize their time to reflect the many demands of their faculty obligations, but they must also prioritize their efforts in ways that will improve the prospects of career advancement. The current perception, as expressed by many in attendance at the workshop conducted as part of this study, is that research contributions are the most important measure with respect to faculty promotion and tenure decisions and that teaching effectiveness is less valued regardless of the stated weighting of research, teaching and service. While this perception may be most applicable at research institutions, these same institutions also confer a preponderance of engineering degrees awarded annually. In addition, methods for assessing research accomplishments are well established—even though imperfect, whereas metrics for assessing teaching, learning, and instructional effectiveness are not as well defined or well established.

In 2007, with support from the National Science Foundation, the National Academy of Engineering convened a committee of engineering educators, leaders in faculty professional development, and experts in teaching assessment to organize a fact-finding workshop and prepare a succinct consensus report that addresses the development and implementation of a system to measure the instructional effectiveness of engineering faculty members. The charge to the committee was to identify and assess options for evaluating scholarly teaching (referred to in this report as "teaching effectiveness") which includes a variety of actions and knowledge related to faculty members' content expertise, instructional design skills, delivery skills, understanding of outcomes assessment, and course management skills. The intent of this project was to provide a concise description of a process to develop and institute a valid and acceptable means of measuring teaching effectiveness in order to foster greater acceptance and rewards for faculty efforts to improve their performance of the teaching role that makes up a part of their faculty responsibility. Although the focus of this report is in the area of engineering, the concepts and approaches are applicable to all fields in higher education.

The study process included a fact-finding workshop that convened 25 experts in the areas of engineering education, institutional administration, and teaching and learning assessment at which three commissioned papers were presented relating to research in assessing instructional effectiveness, metrics that are currently available, and what constitutes effective teaching. Drawing on the commissioned papers, workshop discussions, and additional background research, the committee with support of NAE professional staff prepared a report that addressed the following topics:

- Background, Framing and Concepts
- Governing Principles of Good Metrics
- The Committee's Key Assumptions in Approaching the Task
- Attributes That Should be Measured and Sources of Data
- How to Measure and Compute Teaching Performance

The committee reached the following stipulations and recommendations for action by institutional leaders and external stakeholders of the engineering educational system.

Stipulations

- Faculty enrichment programs on campus often have high enrollments and are sometimes oversubscribed (relative to the resources available to faculty development programs). However, the optional nature of such programs and limited resources leads to low and uneven overall participation.
- The development of a thoughtfully designed and agreed-upon method of evaluating teaching effectiveness—based on research on effective teaching and learning—would provide administrators and faculty members the ability to use quantitative[1] metrics in the promotion and tenure process.
- Quantitative and broad metrics would provide faculty members with an incentive to invest time and effort to enhance their instructional skills.
- All faculty and administrators should have significant input into the design of an evaluation/assessment system, as well as provide feedback based upon the results stemming from the evaluation system that is developed.
- The assumptions, principles, and expected outcomes of assessing teaching effectiveness should be explicit (and repeated frequently) to those subject to the evaluations, as well as to those who will conduct the evaluations.
- Information gathered for tenure and promotion evaluations will likely overlap with information gathered for professional development. However, these two functions should remain separate such that identifying weaknesses for professional development efforts (collecting formative assessment data) is not seen as having potentially negative impacts on tenure and promotion evaluation (summative assessment data). This is a necessary safeguard that maintains faculty members' confidence that sincere effort to improve their teaching through honest evaluations of strengths and weaknesses will not result in downgraded tenure and promotion evaluations.

Recommendations

Institutions, engineering deans and department heads should:

- Use multidimensional metrics that draw upon different constituencies to evaluate the content, organization, and delivery of course material and the assessment of student learning.
- Take the lead in gaining widespread acceptance of metrics for evaluating teaching effectiveness in engineering. Their links to faculty and institutional administrators give

[1] The use of the word *quantitative* with respect to the proposed approach implies that the broad set of metrics that can be adopted are then given a numeric value whether the data are derived from sources that are quantitative or qualitative in nature. For example, an assessment of delivery skills is clearly a qualitative assessment; however, rating delivery skills on a scale of 1 to 4 creates an assessment that can be used quantitatively as part of a larger evaluative system.

them the authority to engage in meaningful dialogue in the college of engineering and throughout the larger institution.

- Seek to develop the appropriate number of evaluators who have the knowledge, skills, and experience to provide rigorous, meaningful assessments of instructional effectiveness (in much the same way that those institutions seek to ensure the development of the skills and knowledge required for excellent disciplinary research).
- Seek out and take advantage of external resources, such as associations, societies, and/or programs focused on teaching excellence (e.g., Carnegie Academy for the Scholarship of Teaching and Learning, Higher Education Academy [U.K.], and Professional and Organizational Development Network), as well as on-campus teaching and learning resource centers and organizations focused on engineering education (e.g., the International Society for Engineering Education [IGIP][2] and the Foundation Engineering Education Coalition's web site devoted to Active/Cooperative Learning: Best Practices in Engineering Education http://clte.asu.edu/active/main.htm).

Leaders of the engineering profession (including the National Academy of Engineering, American Society for Engineering Education, ABET, Inc., American Association of Engineering Societies, the Engineering Deans' Council, and the various engineering disciplinary societies) should:

- Continue to promote programs and provide support for individuals and institutions pursuing efforts to accelerate the development and implementation of metrics for evaluating instructional effectiveness.
- Seek to create and nurture models of metrics for evaluating instructional effectiveness. Each institution, of course, will have particular needs and demands; however, nationally known examples of well informed, well supported, and carefully developed instructional evaluation programs will benefit the entire field.

[2] The group's acronym, IGIP, is attributable to its name in German, "Internationale Gesellschaft für Ingenieurpädagogik."

1
Background, Framing, and Concepts

In this report, we present a case for using metrics to evaluate teaching and other kinds of instruction in engineering, based on the premise that the quality of their teaching should be a factor in decisions about faculty appointments, promotions, and advancement. This will be possible only if metrics for evaluations are widely available, easy to use, and recognized and respected by all. Metrics for evaluating teaching and learning are also important for ensuring that students receive the best possible education. In addition, meaningful evaluations provide all faculty members with opportunities for continuing their professional development and improving their teaching knowledge and skills.

Practical factors also support the need for evaluating teaching and learning. State budgets for higher education have become increasingly constrained and subject to competition from other pressing needs. This long-term, sustained trend has led to more rigorous scrutiny of public institutions of higher education by state governments and agencies, as well as agencies at the federal level, as reflected in *A Test of Leadership: Charting the Future of U.S. Higher Education*, more frequently called the "Spellings Commission Report" (United States Department of Education, 2006). A natural result of this oversight movement is a growing demand for accountability in higher education (i.e., the ability to show that public dollars are being used wisely and are engendering tangible, positive results). The primary area of accountability is teaching, or more correctly learning. Do students receive an educational benefit that justifies the dollars they and their families spend, and does society receive an appropriate benefit in return for the public dollars spent on education? The capability to document the quality of teaching and learning is a necessary part of institutional accountability, and the documentation methods used must be understandable, transparent, and cogent.

The ability of faculty to promote student learning is a natural consequence of their preparation for their role as teachers. In the past few decades, some institutions have established future-faculty programs for graduate students and/or teaching-resource centers for current faculty. In addition, the National Science Foundation (NSF) and other groups and organizations have supported numerous projects and centers focused on teaching and learning. A listing of some of the various centers and coalitions that are concerned with engineering and are located throughout the United States is given in Figure 1.1 (Atman, 2007). Although these programs are usually voluntary and their success varies, faculty-development centers have generated moderate to high interest among engineering educators (Van Note and Szabo, 1996).

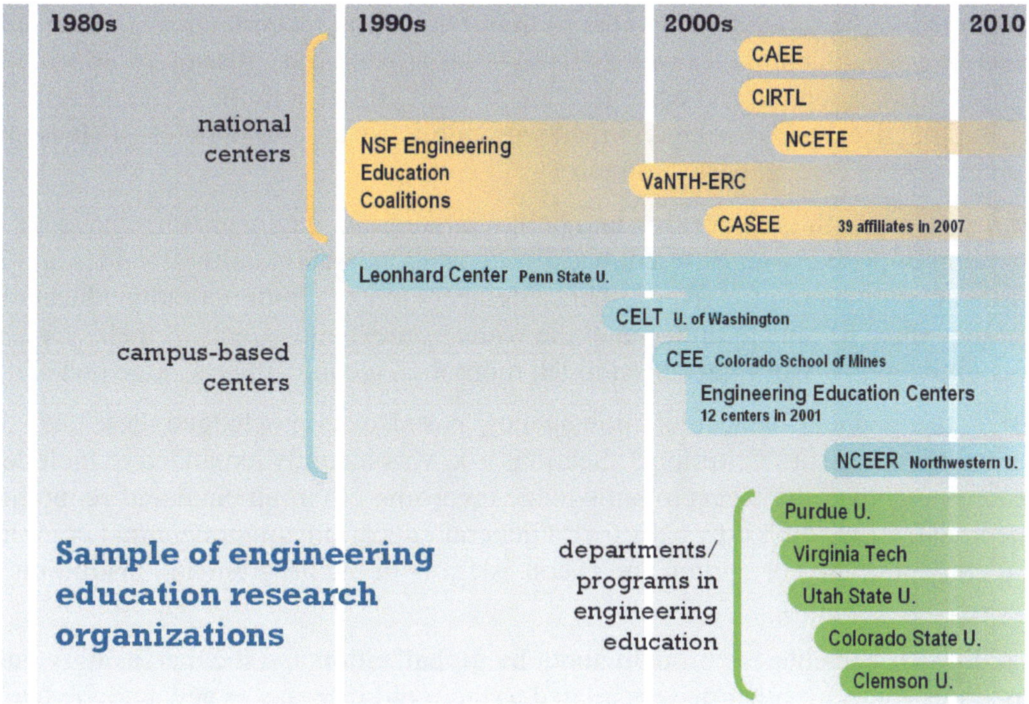

Figure 1.1 Engineering Education Organizations and Centers (adapted from Atman, 2007).

To a large extent, however, graduate students who plan to pursue academic careers receive little or no supervised instruction in teaching. Thus new faculty members must usually develop their own approaches and styles of teaching. The application of metrics and institutional resources for evaluating progress would make it much easier for all faculty members to develop and improve their teaching skills continually. Additionally, a formal evaluation of teaching effectiveness would likely benefit graduate students' exposure to teaching and learning concerns as well as their preparation for future teaching responsibilities.

Another reason for evaluating teaching and learning is that the demands on practicing engineers are changing, and the system for engineering education must necessarily change with those demands. In this fast-moving environment, it is important to assess how teaching is changing and whether the changes are effective.[1]

DRIVERS OF CHANGE IN ENGINEERING EDUCATION

The need for effective evaluation of teaching is an ongoing one; however, there are a number of recent developments that impact upon engineering education and make evaluation all

[1] Concerns about the quality and efficacy of higher education have elicited a variety of responses. Two prominent examples are a workshop in the Association of American Medical Schools, *Advancing Educators and Education: Defining the Components and Evidence of Educational Scholarship* (2007). Washington, D.C.: AAMC, and a review of indicators of quality teaching and learning by the Carrick Institute (Australia) in Chalmers. D., *A Review of Australian and International Quality Systems and Indicators of Learning and Teaching* (2007). Chippendale, NSW: The Carrick Institute for Learning and Teaching in Higher Education, Ltd. [Note: The Carrick Institute is now known as the Australian Learning and Teaching Center.]

the more critical. The rapid development of high-bandwidth technologies has enabled instantaneous communication, as well as rapid access to and transmission of information. Industry, business, research, and education are now all global activities. In addition, engineering projects and practices today are intertwined with public issues and policies, such as energy, the environment, health care, and government.

For reasons of both globalization and public interaction, engineers must have an understanding of people with different backgrounds and different cultural values and must be able to interact with them effectively. Thus engineers must be more broadly educated than in the past, and they must be able to understand the wider context and effects of their work. Successful engineers in the global workplace need much more than technical knowledge and skills.[2]

ABET, the engineering accreditation agency, has also acknowledged these changes in the engineering environment. Criterion 3, Sections a-k, was recently expanded to include elements of increased breadth with respect to knowledge regarding environmental and economic assessment, and Criterion 5 now calls for "a general education component that complements the technical content of the curriculum and is consistent with the program and institution objectives"(ABET, 2008).

In addition to the changes brought about by globalization and the increasingly public nature of engineering, there are other drivers related to the rapid increase in new knowledge, the changing characteristics and capabilities of students, and the utilization of engineers in the workplace.

- The knowledge base is increasing exponentially, and with new technologies, this knowledge is readily retrievable through digital libraries and the Internet. We have long since passed the point at which all of the knowledge necessary for a career in engineering can be packed into the undergraduate engineering curriculum. Students today must be taught to be lifelong learners able to think critically and be adept at locating, evaluating, and assimilating information from many sources.
- Students entering college today have grown up in an information society enriched by technological capabilities, an environment in which instant communication, information searching, and multi-tasking are routine.[3] Engineering educators must recognize these capabilities, make use of them to advance learning, and build upon them when appropriate.
- A vast majority of engineering graduates, 94 percent by some estimates, pursue careers in engineering practice rather than in academia (NSF, 2002). However, educators at research universities have a natural tendency to emphasize fundamental knowledge and research methodologies in their courses rather than practical experience and engineering practice. To meet the needs of today's students, the authoring committee believes that increased focus should be given to disciplinary aspects of engineering practice and to

---

[2] These issues are discussed in detail in two recent reports from the National Academy of Engineering, *The Engineer of 2020: Visions of Engineering in the New Century* (2004) and *Educating the Engineer of 2020: Adapting Engineering Education to the New Century* (2005) (Washington, D.C.: The National Academies Press), and a report by the University of Michigan Millennium Project, *Engineering for a Changing World.*

[3] In some studies, multi-tasking has been shown to interfere with the learning of abstract and complicated material. (Foerde, K., Knowlton, B., and Poldrack, R. (2006). Modulation of competing memory systems by distraction, in Proceedings of the National Academy of Sciences, vol. 103 no. 31, 11778-11783. Washington, DC.)

teaching these aspects explicitly. One of the conclusions of *Educating the Engineer of 2020* stated that we need to put more emphasis upon engineering practice in university engineering education programs, and movements in this direction will impact on both who is teaching as well as how we teach (NAE, 2005).

IMPROVING METHODS OF EVALUATION

Most college faculty members have had little or no formal training in the complex and intellectually sophisticated skills necessary for designing and delivering instruction or in assessing student learning outcomes (Felder, 1993; Brint, 2008), Therefore, effective teaching requires that educators adapt, develop, and hone their teaching skills to increase the level of student learning, and institutions must provide the resources to help them to develop, support and improve their teaching performance—especially if teaching performance is to be evaluated. The US Department of Education, has emphasized the importance of assessing learning outcomes as part of the accreditation process (which mirrors the intent of changes to engineering accreditation efforts in the ABET EC2000 criteria). This emphasis has been expressed in a number of accreditation standards including standards that refer specifically to faculty evaluation (see, for example, Comprehensive Standard 3.7.2 of the Southern Association of Colleges and Schools accrediting agency). Thus, as the accreditation standards for higher education begin to focus more on learning outcomes and the assessment procedure used to ensure quality teaching, the development and use of a consistent set of quality metrics for assessing faculty teaching becomes increasingly important in all aspects of higher education. An informative resource for efforts to identify the component parts of teaching that should be included in a matrix of skills and attributes is the National Research Council report called *Evaluating and Improving Undergraduate Teaching in Science, Technology and Mathematics.* The report provides an excellent overview of the area of teaching effectiveness (NRC, 2003).

Practical realities must be taken into consideration in the development of metrics for faculty evaluations. University faculty members have very full schedules and many commitments inside and outside the classroom. Because change requires an investment of time, there is a strong tendency among them to maintain the status quo. Thus the methods and metrics for evaluating teaching and learning must be efficient in terms of how much time they require, and they must be user friendly.

Current assessment methods are heavily dependent on student ratings, which may provide only a single dimension of the classroom experience. Effective metrics should also include diverse and complementary methods. Information technology can support the documentation of faculty activities that reflect their growth and development with respect to improved teaching and learning efforts.

As professionals, most faculty members value self-governance and self-policing. Therefore, for change to be effective, they must "buy in" to the need for a systematic approach for measuring teaching effectiveness. Buy-in is more likely if faculty members have the opportunity to participate in the identification and implementation of methodologies through, for example, academic senates and other activities. They should also be involved in determining how assessments by other faculty members will be used in making decisions and evaluating their skills.

Finally, the leaders of a profession define the values and priorities of that profession, and their support carries considerable weight. The National Academy of Engineering (NAE), engineering societies, and other stakeholders can provide valuable support by emphasizing the benefits of evaluating teaching and learning. Leaders of NAE and other societies should put forward statements indicating the need for quality teaching that can transform students from vessels of knowledge into sophisticated seekers and users of knowledge.

2

Governing Principles of Good Metrics

Important first steps in creating metrics for evaluating teaching in engineering schools is to develop principles that ensure that the metrics will be widely accepted and sustainable and that they actually will provide valid assessments of the educational impact of faculty on students. One of the main principles should be *what is valued is rewarded, and what is rewarded is valued.*

For far too long many have bought into the notion that teaching effectiveness cannot be evaluated as objectively as research contributions (where output quantity, frequency of citation, and confidential letters attesting to quality and impacts are frequently employed; England, 1996). Some have internalized this principle and made it part of the value system in engineering education, namely, that teaching is less important and less scholarly than research. Promoters of metrics for evaluating teaching must be sensitive to these long-held, very strong convictions and recognize that introducing metrics will represent a major cultural change.

The principles listed below are common to the development of any new system in an organization and can guide the creation of metrics for evaluating teaching:

- *The evaluation system must be compatible with the overall mission, goals, and structure of the institution* because engineering colleges reside within universities, and the evaluation of engineering faculty for promotion and tenure will eventually be conducted by university committees. If metrics have been created in isolation, engineering faculty might be judged by one set of criteria in the engineering context and a different set of criteria in the context of promotion at the university level. Thus, ideally, engineering schools should approach their respective institutions to initiate a discussion across the university regarding improved metrics for evaluating teaching.

- *The proper locus for developing an effective evaluation system should be the deans and department chairs, or their equivalents.* These administrative levels can provide the necessary connections between the institutional administration and the individual faculty members. Deans and department heads can also assist in allocating resources for the design and implementation of an evaluation system that is in concert with the institutional mission, goals, and structure.

- *To ensure the acceptance of the evaluation system, faculty members should be integrally involved in its creation* (i.e., faculty must believe in the fairness and utility of the evaluation process). To ensure faculty buy-in, they must be involved in the discussions from the beginning. Moreover, the discussions themselves, by providing a forum where faculty from different departments can discuss characteristics and methods of effective teaching, will begin to break down the barriers of teaching as an isolated activity and reposition it as a collegial activity, thus further legitimizing its value.

- *The evaluation system should reflect the complexity of teaching, which must include the course design element, implementation and delivery of the course, assessment, and mechanisms for continuous improvement, and recognition of different learning styles and levels of student abilities.* Teaching is both a science and an art, and doing it well requires a knowledge base and skills that are usually not well-addressed in disciplinary doctoral programs.

- *At the end of the day, the discussion participants must be in agreement/consensus on the fundamental elements of effective teaching.* Most important, learning[1] should be a key component of any definition, because the outcome of effective teaching is always learning. Other elements include design (e.g., the alignment of clearly articulated objectives/outcomes,[2] assessments,[3] and instructional activities[4]) and implementation (e.g., clear explanations, frequent and constructive feedback, illustrative examples).

- *An evaluation of teaching should include both formative feedback to assist/help individual improvement and summative evaluation to measure progress toward institutional goals.*[5] An evaluation system must identify areas for improvement and provide both opportunities and support for making those improvements. While we believe that faculty evaluation and faculty development should not be programmatically linked (they should not be housed in the same entity or done by the same people), linking the two conceptually sends a clear message that the institution supports faculty growth, which happens only when faculty receive ongoing and constructive feedback.

- *The evaluation system must be flexible enough to encompass various institutional missions, disciplines, audiences, goals, teaching methodologies, etc.* In addition, it should also accommodate people on different "tracks" (e.g., some universities have adopted teaching tracks as some faculty gravitate toward expanded teaching roles at different points in their careers). Finally, the system should be flexible enough to acknowledge, encourage, and/or reward educational experimentation or attempts at educational innovation. A flexible system enables instructors to try new things without worrying that they might be penalized if the outcomes are not immediately positive.

---

[1] In the context of this report, learning is defined as knowledge, skills, and abilities, as well as attitudes students have acquired by the end of a course or program of study.

[2] Objectives/outcomes are descriptions of what students should be able to do at the end of the course (e.g., analyze, use, apply, critique, construct).

[3] Assessments are tasks that provide feedback to the instructor and the student on the student's level of knowledge and skills. Assessments should be varied, frequent, and relevant.

[4] Instruction includes providing contexts and activities that encourage meaningful engagement by students in learning (e.g., targeted practice).

[5] A formative assessment is typically defined as an ongoing assessment intended to improve performance, in this case, faculty teaching (and hence student learning). A summative assessment, typically conducted at the end of instruction (e.g., of a semester or program), is used to determine overall success.

- *Evaluations should be based on multiple sources of information, multiple methods of gathering data, and information for multiple points in time.*[6] The evidence collected should be reliable (i.e., consistent and accurate), valid (i.e., it should measure what it is intended to measure), and fair (i.e., it should reflect the complexity of the educator's achievements and accomplishments).

- It is equally important to note that *collecting and analyzing data of this sort often demands a skill that we may need to develop further among our faculty and administrators.* A good way to learn these skills might be to enlist the help of colleagues on campus who have expertise in, for example, survey design, qualitative interviewing, educational outcomes research, and so forth.

- *A sustainable evaluation system must not require implementation that is burdensome to faculty or administrators.* However, it is important to guard against sacrificing the fairness, validity, accuracy, and reliability of the evaluation system in trying to make it as easy to use as possible.

- *The evaluation system itself should be evaluated periodically to determine if it is effective.* These periodic reviews should be part of the development plan to ensure that evaluations provide both formative feedback that leads to improvements in teaching and data adequate for judging the quality of teaching.

If the system is successful, all stakeholders will recognize that it provides accurate and valuable information that meets the needs of various groups and creates a culture of assessment that drives teaching and learning improvements. They will also agree that an assessment is not *done to* faculty but is done *by* faculty and *for* faculty and that assessment supports continuous improvements in the quality of education. If stakeholders internalize the principles listed above for developing metrics, they will naturally support a culture of assessment.

[6] Both direct and indirect measures should be used. Direct measures (e.g., exams, projects, assignments) show evidence of students' knowledge and skills. Indirect measures (e.g., teaching evaluations) reflect students' perceptions of teaching effectiveness and employers' and alumni perceptions of how well the program prepares students for their jobs.

3

ASSUMPTIONS

The basic, critical assumption that underlies this report is that a well developed, meaningful mechanism for evaluating instructional effectiveness will improve both teaching and learning. This assumption is based on the common understanding that faculty (like most individuals) respond in accordance with how well their efforts are rewarded. As stated earlier, the perception is that the current system for evaluating faculty for promotion and tenure is heavily weighted in favor of research (scholarly and creative activities) with a relative low weight given to teaching. This imbalance reflects that in "the market" in higher education, effective teaching, unlike research, is not rewarded with advancement and prestige.

Another reason for the imbalance might be that the methods used to evaluate teaching effectiveness are not well developed or widely understood and, in most cases, have not been adopted at the institutional level. Under these circumstances, administrators may be understandably reluctant to give significant weight to an assessment whose validity and accuracy may be uncertain, or even suspect.

Another significant underlying assumption is that all faculty members are capable of improving their teaching. Just as researchers must constantly update their knowledge and methodologies, instructors should also continue to "update" their teaching practices based on both developments in learning and pedagogy and feedback on their teaching skills.

Also, an assumption which is closely linked to the preceding assumption is that *many faculty members are intrinsically motivated to improve their teaching. Therefore, they may welcome feedback, both formative and summative, if it is believed that it will improve their teaching effectiveness.* Of course, the committee is aware that priorities among demands on faculty for research, service, and personal life, as well as teaching, differ among types of universities, from university to university within a type, from department to department, from individual to individual, and even from time to time.

Some people may question whether all, or even most, engineering educators have an intrinsic desire to improve their teaching. Certainly, the responses of some faculty members to teaching evaluations seem to exhibit more cynicism than intrinsic motivation. However, faculty members are typically high achievers and are concerned with how they would be ranked in comparison with their peers being similarly evaluated. Therefore, we assume that when faculty members feel that the information they receive from teaching evaluations is appropriately informative, they will use that information to improve their teaching. Thus the crucial factor is that faculty members must believe that an evaluation system is appropriately informative.

Although it may appear that some faculty would not welcome feedback on their teaching, it is likely they are reacting within the context of current promotion and tenure and evaluation systems. Any performance evaluation must be perceived to be accurate and fair in order for the individuals being evaluated to welcome the experience and to try to improve their performance by changing their teaching practice. Of course, even if a system is perceived to be "unfair," it may still lead to changes in behavior, provided the outcome of the evaluation is sufficiently threatening. However, we are more interested in developing an evaluation system that motivates changes because the system is fair and informative, rather than because it is threatening.

While the issue of accuracy of such instruments is a subject that is broadly understood and does not warrant in-depth description in this text, the issue of fairness will be defined more thoroughly. The perception of fairness cannot be separated from the egocentrism of the person being evaluated. A study by Paese, Lind and Kanfer (1988) found that pre-decision input from those who will be judged in the evaluation process will lead to their judging the system to be procedurally fair. However, many other investigators have demonstrated that, even for those who have had input into developing the process, perceptions of fairness are linked, consciously or not, to an individual's interests and needs (Van Prooijen, 2007). Thus a sense of fairness is significantly affected by whether an individual believes he or she may benefit from an action, or, even more important, whether he or she will be disadvantaged by it. Thus all individuals, even those who had input into the development of a process of evaluation, may eventually or initially consider the system unfair, depending upon how the system influences decisions that affect them.

With respect to implementing a more effective and valuable assessment program, we might adapt to instruction a practice commonly used to increase competence in the evaluation of research proposals and journal articles. That is, we can systematically engage graduate students and junior faculty in evaluating the various types and aspects of teaching effectiveness. Their reviews of teaching are then evaluated by senior faculty as a way of providing valuable feedback and constructive criticism on the quality and comprehensiveness of the reviews. The time and effort of graduate students and junior faculty pay off by raising the level of their understanding of the research, teaching, and reporting process as a whole. At the same time, their efforts ensure that future cadres of effective reviewers and researchers will be available.

Similar efforts could be made to increase competency in instructional evaluation by enlisting senior faculty with expertise in teaching along with the participation of graduate students and junior faculty to increase their capabilities as evaluators of instructional effectiveness. Such an investment would utilize the approach used to foster continuous improvement in research techniques through advising and mentoring of graduate students and junior faculty not only to ensure that more, and more capable, individuals had some experience of assessing instructional effectiveness, but also to create a large cadre of faculty with exposure to the concepts of instructional design and delivery and a better understanding of the fields of instructional research.

Our final assumption is that *administrators and campus reviewers will do their jobs fairly and objectively, including making appropriate assignments, communicating university and program expectations, and using the data collected from evaluations to make fair and accurate judgments of performance, both to encourage professional development and to inform job-advancement decisions.* This assumption assumes a great deal of trust and requires some further explanation.

The ultimate goal of evaluating teaching is to provide feedback to individuals (in both formative and summative formats) as a basis for gauging their effectiveness in meeting institutional and program expectations and then continuously improving their teaching performance to satisfy their intrinsic desire for excellence. To accomplish this goal, the individuals being evaluated must depend upon a team of people to gather and analyze data in a way that they trust will produce accurate and fair results.

As Lencioni (2002) points out, no team can function effectively without trust. In university settings, administrators cannot create an environment of trust by themselves, but they can be crucial players in maintaining trust. Some of the things administrators and campus reviewers should do to engender trust in the teaching evaluation process are listed below:

1. They must assign faculty to teach only in areas in which they have, or can readily develop, the expertise to teach at an appropriate level.

2. They must ensure that an evaluation of an individual's teaching performance is considered in the correct context, such as expected outcomes for student learning, the level of students in the course, whether a course is required or elective, the size of the classes and the nature of the available facilities, and the past experience of the instructor in this teaching situation.

3. Complex social data, such as teaching evaluations, must be used in accordance with well documented social science practices that have established appropriate interpretations and limitations for deriving results.

4. Administrators and reviewers must show that they are using the evaluation process to develop and advance faculty members fairly.

4

What To Measure

The assumptions and governing principles discussed in chapters 2 and 3 provide a framework for developing detailed procedures for determining *what* to measure in evaluating teaching and *how* to measure it. The faculty must be engaged not only in determining *what* to measure, but also in how to "weight" each measure. Thus faculty values and priorities must be taken into account, as well as the mission and goals of the larger institution. Any evaluation system is predicated on a set of values. That is, a set of desirable conditions is defined and then measurements are made to determine whether those conditions have been met. However, the determination as to what constitutes a desirable condition is dependent upon the values held by those interested in developing the evaluation system. Thus in designing a faculty evaluation system the "desirable" conditions to be met must be expressed in terms of the "value" that faculty place on teaching, research productivity, service, and other faculty activities. For example, if research productivity is to be valued more than teaching effectiveness, then a greater weight must be placed on the metric resulting from the measurement of research productivity as compared to the weight placed on the metric resulting from the measurement of teaching performance. Combining the weighted measures of the various faculty roles produces an overall evaluation metric that reflects the "faculty value system" and is thus more likely to be seen by the faculty as being a valid system. The process involves at least four major steps:

1. Define and clarify the underlying terms and assumptions on which the evaluation system is based.

2. Define the value system of the faculty by systematically engaging faculty in defining the following conceptions (which have expanded discussion in later sections of the report):

    - the forms of teaching in engineering education
    - the characteristics (or performance elements) of effective teaching in engineering
    - the value, or "weight" of various characteristics (or performance elements) in the overall evaluation of teaching performance
    - the appropriate sources of information to be included in the evaluation

3. Integrate faculty values and institutional values to ensure that engineering faculty will be able to compete fairly for institutional promotions and tenure.

4. Develop and/or select appropriate tools for measuring the performance elements of effective teaching as determined by the faculty.

The remainder of this chapter describes steps 1 and 2 which address the broad question of *what* to measure. Steps 3 and 4 which relate to *how* to measure, are addressed in Chapter 5.

STEP 1: BASIC TERMS AND UNDERLYING ASSUMPTIONS

The purpose of this step in the development process, which takes place before the faculty become involved, is to define the basic terms, such as *measurement* and *evaluation*, and clarify the underlying assumptions of the evaluation, such as that the goal is to design an evaluation system that will be objective and fair.

Definitions of Terms

In the physical sciences, the term *measurement* is generally defined as *the numerical estimation and expression of the magnitude of one quantity relative to another* (Michell, 1997). However, this definition makes sense only for measuring physical and observable objects or phenomena. When measurement is used in the context of an evaluation of teaching, it takes on a somewhat different meaning, because the "things" being measured do not have readily observable, direct, physical manifestations.

For example, an evaluation that measures the impact of a faculty member's teaching on students' cognitive skills and/or attitudes may be desired. Although there may be some direct external evidence of these, such as student performance on examinations, this measurement will likely involve gathering certain types of data (e.g. student ratings, peer opinion questionnaires) as a basis for *inferring* a measurement of an internal cognitive or affective condition.

The terms measurement and evaluation are not synonymous. A measurement is as objective and reliable as possible. Whereas measurement involves assigning a number to an observable phenomenon according to a rule, evaluation is defined as the interpretation of measurement data by means of a specific value construct to determine the degree to which the data represent a desirable condition (Arreola, 2007). *Thus the result of an evaluation is a judgment, which, by definition, is always subjective.*

A specialized field of psychology, called psychometrics, has been developed to perform the kinds of measurements used in evaluations. Psychometrics is discussed in greater detail in the next chapter on how to measure the performance elements of teaching.

The Assumption of Objectivity

When an institution undertakes to develop a faculty evaluation system, the goal is to ensure that the system is as objective as possible. However, total objectivity in a faculty evaluation system is an illusion, because the term evaluation, by definition, involves judgment, which means that subjectivity is an integral component of the evaluative process.

In fact, the term *objective evaluation* is an oxymoron. Even though the measurement tools used in a faculty evaluation system (e.g., student ratings, peer observation checklists, etc.) may achieve high levels of objectivity, the evaluation process is, by definition, subjective.

However, the underlying rationale for wanting an "objective" faculty evaluation system is to ensure fairness and to reduce or eliminate bias. Ideally, in a fair, unbiased evaluation system anyone examining a set of measurement data will arrive at the same evaluative judgment. In other words, such an evaluation system would produce consistent outcomes in any situation.

Definition of Controlled Subjectivity

Since a completely "objective" evaluation is not possible, however, the goal must be to achieve consistent results from a necessarily subjective process. That is, we must design a process that provides the same evaluative judgment based on a data set, regardless of who considers the data. This can be done through a process called *controlled subjectivity.*

Psychometric methods can be used to create tools for measuring faculty performance (e.g., observation checklists, student- and peer-rating forms) in a way that produces reliable data (i.e., measurements) that are as objective as possible. However, because we know that an evaluation must be subjective, the problem is how to achieve the characteristic of objectivity (i.e., consistency of conclusions based on the same data regardless of who considers them) in a necessarily subjective process.

Because subjectivity in a faculty evaluation system is unavoidable, the goal should be to limit or control its impact. To accomplish this we use a process called *controlled subjectivity,* which is defined as *the consistent application of a predetermined set of values in the interpretation of measurement data to arrive at an evaluative judgment* (Arreola, 2007).

In other words, subjectivity in an evaluation system can be controlled when an *a priori* agreement has been reached on the context and (subjective) value system that will be used to interpret the objective data. Thus, even though the evaluation process involves subjectivity, we can still ensure consistency in outcomes, thus approximating a hypothetical (although oxymoronic) "objective" evaluation system.

STEP 2. DETERMINING THE VALUE SYSTEM

Every evaluation rests upon an implicitly assumed value or set of values. An evaluation provides a systematic observation (measurement) of the performance of interest and a judgment as to whether that performance conforms to the assumed values. If there is a good match, the performance is judged desirable and is generally given a positive or "good" evaluation. If there is a discrepancy, the performance is judged to be undesirable and is generally given a negative or "poor" evaluation.

As was noted earlier, the evaluation process implies the existence and application of a contextual system, or structure, of values associated with the characteristic(s) being measured. Thus before an evaluation system can be developed, the values of those who intend to use it must be defined and should be carefully developed to reflect the values of the institution where they will be applied. For a faculty evaluation system to reflect the values of the institution correctly, we must not only determine those values and have them clearly in mind, but we must also express them in such a way that they may be applied consistently to all individuals subject to the evaluation process.

The "Faculty Role" Model

The value system of a faculty evaluation for a unit in a larger institution must be in basic agreement with the larger value system of the institution. The first step, therefore, must be to ascertain the institution's *"faculty role" model*, that is, the various professional roles faculty are expected to play and how much weight is given to performance in each role in the overall evaluation of the faculty—especially as that evaluation impacts decisions about tenure and promotion.

The *faculty role model*, often described in a faculty handbook or other personnel manual, generally specifies the traditional roles of *teaching, research,* and *service.* Recently, however, many institutions have adopted a more comprehensive faculty role model—*teaching, scholarly and creative activities,* and *service*; in addition, *service* is described in more detail as *service to the institution, the profession,* and *the community.* Whichever faculty role model the institution has adopted must be the starting point in the development of a faculty evaluation system.

In an evaluation system, the institution's mission, goals, priorities, and values may be expressed as "weights" assigned to the performance of each role. Traditionally, the faculty role model was weighted as follows: teaching 40 percent; research 40 percent; and service 20 percent. However, the consensus opinion of workshop participants indicated that faculty often perceive that the "actual weighting" is skewed toward research and does not adhere to the nominal weightings in the model.

Today, many institutions are adopting a more flexible faculty role model in which the research component has been expanded to include scholarly and creative activities (e. g., consulting and practice, generalization and codification of knowledge to give deeper insights, serving on national boards and agencies, translating basic research results into practical products or services, and even creative new approaches to education), and the weights have been adjusted to reflect the complexity of faculty work assignments. Thus some current faculty role models may look more like the one shown in Table 4.1.

TABLE 4.1 Faculty Role Model with Value Ranges

Minimum Weight	Faculty Responsibilities	Maximum Weight
20%	Teaching	60%
30%	Scholarly/Creative Activities	70%
10%	Service	15%

As Table 4.1 shows, research has been redefined as scholarly/creative activities, and the weights are expressed as ranges rather than fixed values. In this example, the weight assigned to teaching in the evaluation ranges from 20 percent to 60 percent. The range-of-values approach is useful in that it reflects the diversity of faculty assignments in the institution, or even in a single department.

An instructional unit must base its faculty evaluation system on whichever type of faculty role model the institution has adopted. Thus, if the model includes ranges, the unit must weight its evaluation of teaching in a way that corresponds to, or falls within, the ranges adopted by the institution. In short, the faculty evaluation system of the unit must adhere to the *governing principle* described in Chapter 1, of being compatible with the mission, goals, and values of the larger institution.

In the event that an institution has not adopted a faculty role model that specifies weights or weight ranges, a unit might develop its own weighting scheme. The unit might then be in a position to take the lead in working with the institutional administration to clarify the values, and thus the operational weights, for evaluations of faculty for determining promotions and tenure.

Faculty Participation

Faculty must be systematically involved in determining and defining the faculty role model as it relates to the institutional mission and values since this process is a necessary first step. Because the evaluation of teaching requires *gathering* various measures and then *interpreting* them by means of *a value construct*, determining and specifying the institutional values is a continuous process. Although it is advisable to establish a coordinating committee or task force to carry out this process, it is also critical that the larger faculty be engaged in the discussions to determine their values about the professional execution of their teaching roles.

Faculty may be engaged in many ways. One that has been found to be effective is by scheduling a series of dedicated departmental or college faculty meetings in which faculty members are asked to discuss and come to a consensus about the following issues:

- Agreement on a value, or range of values, assigned to the teaching role in the overall evaluation of a faculty member. Even if values are already specified in the institution's faculty role model, it is important that the engineering faculty clarify the value system for engineering in terms of its congruence (or non-congruence) with the institutional faculty value system.

    - The result might be expressed in a statement similar to the following example**:** In the College of Engineering, the weight assigned to teaching in the faculty evaluation system must reflect the type and amount of teaching a faculty member is required to do in a given academic year and may take on a value within the range of 20 percent to 60 percent in the overall evaluation.

- Agreement on a list of types of teaching situations that should be included in the evaluation (e.g., standard classroom teaching, large lectures, online teaching, laboratory teaching, project courses, and/or mentoring).

    - The result might be expressed in a statement similar to the following example: When one is evaluating teaching, only data from the following teaching environments shall be considered: standard classroom teaching: large lectures, laboratory courses, online courses, project courses, and assigned mentoring. Mentoring graduate student research, which can be categorized as "creative or scholarly activity," and serving as an advisor to student organizations, which can be categorized as "service," shall not be considered evidence of teaching effectiveness for the purposes of a formal evaluation.

- Agree on the characteristics or performance elements (e.g., organization of material, clarity in lecturing, timely replies to e-mail in teaching online courses) that faculty consider necessary for teaching excellence in each type of teaching situation.

    - The result of this effort might be expressed in a substantial report. The underlying problem in the evaluation of teaching has been that the professoriate has not reached a consensus on a definition of what constitutes an excellent teacher. Although considerable research on teacher characteristics and performances that positively influence learning has been done, no universally accepted definition or list of qualities can be found in the lexicon of higher education. If there were such a definition or list, the evaluation of teaching would be relatively easy.

        Many faculty members and academic administrators consider the main component of teaching excellence to be content expertise. Others argue that teaching excellence is an ephemeral characteristic that cannot be measured but results in long-term, positive effects on student lives, of which the instructor may never be aware. The differences between these two opinions (and many others) may never be resolved to everyone's satisfaction.

        Nevertheless, the process of designing an effective learning experience is, to some extent, familiar to engineers, who are adept, or at least familiar, with design processes and the iterations necessary to deliver a product. Designing and delivering an excellent course or learning experience can be thought of in much the same way.

        First, he or she must identify the requirements (e.g., the learning outcomes for the course, what the student needs for learning are, what the profession defines as competencies in knowledge and skills). The instructor must have sufficient expertise in the disciplinary content, as well as in the learning process, to ensure that all students learn. He or she must also establish and refine learning outcomes for students and create learning experiences that are likely to achieve the desired results.

        Once the instructor has designed the course, he or she must deliver the course (i.e., implement the design) and continually evaluate not only student learning outcomes, but also the success of the design. A well designed course may not have the desired effects if other components (e.g., course management) are not handled well. Like all engineering designs, the evaluation of an engineer's work requires input from both customers (i.e., students) and experts in the field (e.g., peers).

- Agree on the most qualified or appropriate sources of information on various characteristics or performance elements in each teaching situation and specify how much weight should be placed on that information.

    - The result of this should be the identification of multiple data sources. At the very least, data from students, peers, and department chairs (or other supervisors) should have input into an evaluation. However, it is important to determine which of these (or other) sources should provide information on the performance of specific elements of teaching in each identified environment, as well as how that information should be weighted.

Table 4.2 shows an example how a faculty member might determine sources of information and how those data sources should be weighted. In this example, input from students counts for 25 percent, from peers 45 percent, from the department chair or supervisor 20 percent, and from the subject of the evaluation 10 percent. The "X's" indicate the appropriate performance elements for which each source should provide information; cells highlighted in gray indicate that no data are to be gathered. The table also indicates the previously determined range (20 percent to 60 percent) for weighting teaching in the overall faculty evaluation.

TABLE 4.2 Example of Data Sources and Weights

Minimum 20%	**TEACHING**		**Maximum 60%**	
Performance Component[1]	**Sources of Measurement Data**			
	Students (25%)	**Peers (45%)**	**Department. Chair/ Supervisor (20%)**	**Self (10%)**
Content expertise[2]		X		X
Instructional design[3]	X	X		X
Instructional delivery[4]	X	X		X
Instructional assessment[5]	X	X	X	X
Course management[6]			X	X

[1] The performance components addressed in this table are commonly discussed topics. Additional source material that discusses these items can be found in the following report: National Research Council. 1999. *How People Learn: Brain, Mind, Experience and School*, Washington, DC.: National Academy Press.

[2] Instructors must be knowledgeable in their specific fields of engineering. However, considerable research has shown that content expertise, although *necessary*, is not *sufficient* to ensure teaching excellence. The concept of pedagogical content knowledge [as described by Shulman, L. (1987). *Knowledge and teaching: Foundations of the new reform.* Harvard Educational Review,_57, 1-22.] describes the connection between discipline content knowledge and pedagogic knowledge that leads to improved teaching and learning.

[3] Instructional design requires planning a logical, organized course that aligns objectives/outcomes, learning experiences (content and delivery), and assessments based on sound principles from the learning sciences.

[4] For effective delivery (implementation), the instructor must use a variety of methods, activities, and contexts to achieve a robust understanding of material, as well as relevant, varied examples of the material and activities that provide meaningful engagement and practice, all of which are aligned with outcomes and assessment methods.

[5] Assessment requires that the instructor design and use valid, reliable methods of (1) measuring student learning of the established objectives and (2) providing meaningful feedback to students.

[6] Course management is judged on how well the learning environment is configured, including equipment, resources, scheduling, and procedures necessary to student learning.

Note that the decisions, made in consultation with faculty, may be entirely subjective. Nevertheless, because this value system will remain constant for all faculty members whose teaching is being evaluated, the subjectivity will be controlled, thus guaranteeing the consistency and comparability of outcomes.

Lengthy discussions and vigorous debate may be necessary for faculty to come to agreement on these parameters. However, agreement is necessary for faculty to feel confident that the evaluation system reflects and respects their conception of excellence in teaching as well as their values and priorities in evaluating teaching. Once the tasks listed in this section have been completed, the process can move to the next stage—determining *how* to measure the performance elements of teaching and how to combine these measures into an overall evaluation of teaching.

5

Measuring Teaching Performance

Up to this point, engaging faculty in the development of the value system; defining the fundamental elements of teaching excellence in engineering education; determining appropriate sources of information in the evaluation of teaching; and weighting the information from these sources have been addressed in operational terms. In this chapter, the subject changes to *how* information should be gathered, assembled, measured, and used, both as part of the institution's reward system and to improve teaching performance.

MEASURING PERFORMANCE ELEMENTS

As noted earlier, content expertise, although necessary, does not guarantee effective teaching. Faculty must be able to design and deliver instructional experiences in such a way that there is some assurance that learning will occur when students engage the experience. The subject matter must be presented in a way that piques students' interest and encourages them to learn. Also, the course design and implementation must provide students with meaningful feedback on their progress in mastering the material.

In addition, teachers must handle myriad routine tasks involved in managing a course. Laboratory supplies must be ordered and inventories maintained, arrangements for guest lecturers must be made, library materials must be put on reserve, field trips must be arranged and coordinated, drop/add slips and, later, grades must be turned in on time, and so on.

Thus effective teaching has many components. Instructors must interact with students in a way that (1) provides opportunities for them to learn; (2) creates conditions that support and facilitate learning; and (3) uses techniques and methods that create an environment with a high probability that students will learn.

At least five basic skills are necessary for effective teaching (Arreola, Theall, & Aleamoni, 2003):

- content expertise
- instructional design skills
- instructional delivery skills
- instructional assessment skills
- course management skills

These are the five performance components that were outlined in Table 4.2.

When the total "act" of teaching is defined in terms of these five broad components, it becomes clear that the evaluation of teaching cannot be accomplished by using a single measurement tool or by basing it on the judgment of one administrator or peer committee who have made a few classroom visits. No one person or group has a detailed, complete view of the entire teaching process.

A more accurate and more valid assessment of teaching performance of necessity involves gathering information on all five dimensions of teaching performance. This might include (1) students' *perceptions* and *reactions* to various aspects of the instructor's delivery, course design, and assessment methods; (2) information from peers, and perhaps informed experts, on the quality of the instructor's design and assessment skills; (3) information from peers and department heads or supervisors on content expertise (primarily in terms of the level, currency, and appropriateness of the material in the course design and supporting materials); and (4) information from the department head or supervisor on the instructor's course management.

Data provided by students would most likely be gathered by a well designed student-rating form that elicits students' perceptions of the effectiveness of the instructional design, delivery, and assessment aspects of the course. Data provided by peers may include reviews of the course syllabus to judge whether (1) the content is current, (2) the design includes experiences that will advance students' mastery of the material, (3) the delivery mechanism (e.g., slides, web pages, lectures, etc.) are well executed, and (4) the assessment tools and procedures are valid and reliable. It should be pointed out that the peers used for such evaluation activities should be experienced and capable of making the assessments that are being asked of them. This will require individuals that have some level of knowledge and expertise in instructional practice.

Data provided by the department chair or supervisor may include (1) external evidence of the content expertise of the instructor, (2) evidence that the instructor is complying with all instructional assessment policies and procedures, and (3) evidence that the instructor complies with internal policies and procedures (e.g., reporting grades, keeping attendance records, supervising laboratory activities, etc.).

Finally, the instructor himself/herself may maintain a portfolio of evidence and/or informal or qualitative evidence on all aspects of teaching performance. Although peers and the department head or supervisor may want to use the portfolio to augment their interpretation, we do not recommend that self-rating data be used in combination with data from other sources, because self-rating data may then have a greater impact than intended. However, determining how much self-rating data should "count" is an issue that should have been resolved at the faculty-engagement stage (Chapter 4).

The key to an effective evaluation of teaching is putting the parts of this mosaic together in a way that accurately reflects the instructor's overall teaching competence.

A DATA-GATHERING RUBRIC

Different units may decide to measure only a subset of the performance components of teaching. Table 5.1, an expanded version of Table 4.2, provides a rubric for gathering measurement data on all of the components of teaching performance. In Table 5.1, the type and source of data is described within the appropriate cell.

TABLE 5.1 Sources and Weights of Measurement Data

Minimum: 20%	TEACHING			Maximum: 60%
Performance Component	**Sources of Measurement Data**			
	Students (25%)	**Peers (45%)**	**Dept. Chair/ Supervisor (20%)**	**Self (10%)**
Content expertise		Review of education, scholarship, professional society activities, assessment of currency in field		Portfolio: evidence of ongoing proficiency in the field
Instructional design	Student rating form	Peer review of course materials (syllabus, readings, experiments, examinations, handouts, etc.)		Portfolio: evidence of a strategy for designing instructional experiences
Instructional delivery	Student rating form	Peer review of course materials (to include items previously listed) combined with peer assessment of classroom presentation skills		Portfolio: evidence of strategies and methods of delivery and communication
Instructional assessment	Student rating form	Peer review of course materials (syllabus, readings, experiments, examinations, handouts, etc.)	Compliance with policies and procedures concerning testing and grading	Portfolio: evidence of techniques, strategies, and methods of assessing student learning and providing meaningful feedback.
Course management			Timely ordering of lab supplies, submission of grades, drop-add slips, etc.	Portfolio: specification of conditions affecting management of the teaching environment.

Tables 5.2 through 5.5 provide strategies for gathering data from various sources.

TABLE 5.2 Strategy for Student Ratings

Description: Students rate an instructor's performance using a structured questionnaire, unstructured questionnaire or interview.
Strengths: Produces extremely reliable, valid information on faculty classroom performance, because students observe the teacher every day (Aleamoni, 1981). Instructors are often motivated to change their behavior as a result of student feedback. If a *professionally designed* student-rating form is used, results show a high correlation with ratings by peers and supervisors; in addition, these assessments are not affected by grades.
Weaknesses: If a professionally developed form is not used, external factors, such as class size and gender, may influence student ratings. In addition, students tend to be generous in their ratings.
Conditions for Effective Use: Student anonymity and instructor's willingness to accept student feedback. Instruments must be carefully developed by appropriate and documented reliability and validity studies.
Nature of the Evidence: Student perceptions of organization, difficulty, and course impact (e.g., how they have changed as a result of taking the course); how various teaching techniques affect them; reactions to instructor's actions; what students like and dislike about an instructor.

TABLE 5.3 Strategy for Peer Ratings

Description: Other faculty or peers rate an instructor's performance in terms of (1) course design, (2) appropriateness and effectiveness of instructional materials, and (3) appropriateness of instructional assessment strategies and tools. Peer reviewers are usually from outside the university, but may include some faculty from within the university. This process would be analogous to peer evaluation as done for research contributions.
Strengths: Raters are familiar with the institutional, departmental, and division goals, priorities, and values, as well as the specific problems that affect teaching. Peer review encourages professional behavior (e.g., a desire to improve one's own profession). Raters with expertise in the instructor's subject area may be able to give content-specific suggestions and recommendations.
Weaknesses: Assumes that peers have expertise in instructional design, delivery, and assessment. Bias may be introduced because of previous personal knowledge, personal relationships, or personal pressure to influence the evaluation. Relationships among peers may suffer. Possible bias may be introduced because of a reviewer's preference for his/her teaching method.
Conditions for Effective Use: A high degree of professional ethics and objectivity. Multiple reviewers.
Nature of the Evidence: Comments on relations between instructor's actions and students' behavior. Comparisons with instructional methods peers may consider superior or more appropriate. Suggestions for instructors on methods to use, etc.

TABLE 5.4 Strategy for Review by Department Head or Supervisor

Description: Administrator evaluates instructor's performance relative to policies and procedures of the colleges and the objectives of the department.
Strengths: Evaluators familiar with college and community goals, priorities, and values often provide additional insights because they can compare the instructor's performance with other performances in the college, school, division, or department.
Weaknesses: Bias may be introduced because of extraneous data, personal relationships, and evaluator's values and favored teaching methods.
Conditions for Effective Use: Requires knowledge of institutional, college, and departmental policies and procedures as they relate to teaching courses in the engineering curriculum and the maintenance of student information (e.g., FERPA, approved grading scale, etc.). Requires maintenance of records relating to instructor's compliance with relevant policies and regulations.
Nature of Evidence Produced: Comments on the relationship between instructor's actions and the achievement of departmental goals and objectives.

TABLE 5.5 Strategy for Self-rating (Portfolio)

Description: Instructor gathers information to assess his/her own performance relative to personal needs, goals, and objectives.
Strengths: May be part of a program of continuous assessment. Likely that instructors will act on data they collect themselves. Data are closely related to personal goals and needs. Necessary to facilitate review of syllabus by peers.
Weaknesses: Results may be inconsistent with ratings by others. Possible unwillingness to collect and/or consider data relative to one's own performance. Tendency to rate performance higher than students do.
Conditions for Effective Use: Requires that instructor be self-confident and secure and have the skills to identifying goals and collect appropriate data. Data cannot be heavily weighted in personnel decisions (e.g., promotion, tenure, merit pay, etc.)
Nature of Evidence Produced: Information on progress toward personal goals.

MEASUREMENT TOOLS

Constructing valid and reliable forms, questionnaires, or other tools for gathering data is a complex task that requires expertise in psychometrics. *We must always keep in mind that what are being developed are tools to measure—in a valid and reliable way—complex psychological phenomena such as opinions, reactions, observations, rankings, and so on.* Even selecting appropriate forms and tools from published, commercially available products requires fairly sophisticated psychometric skills; however, resources to assist in locating instruments can often be found on campus (in educational development office or within the social sciences departments). Each of these products must be assessed for appropriateness and utility in the faculty evaluation system that has been designed for a specific situation.

No standardized forms for peer or department chair ratings are commercially available; however, a search of the internet provides ad hoc checklists, rating forms, and other resources that would provide useful guidance in constructing such tools (University of Texas). Therefore, institutions may have to develop their own—or find appropriate forms that have been used at other institutions. Before either of these can be done, however, it is imperative that the performance elements to be measured have been clearly and completely specified. If new forms must be developed, experts in psychometrics should be consulted. Also, training for the observers is important in that it helps to focus their observations around the items listed on checklist or rating forms (Braskamp & Ory, 1994). Such expertise may be available in other colleges on the campus, especially in departments that focus on educational research, instructional-systems design, or psychological measurement.

All of the tools for the evaluation of teaching must use the same scale of measurement.

That is, whether data are gathered via a student rating form, a peer review form, or a department chair review form, all measures must be on a common scale. Most student rating forms use either a 4-point or 5-point scale. Thus student ratings are represented by a number between 1 and 5, with, in most cases, the highest number indicating the most positive rating. If that scale is adopted, the forms used to gather information from all sources should use the same number scale in reporting results.

COMPUTING AN OVERALL EVALUATION

Once measurement tools have been selected and/or developed for all input sources (Table 3), the systematic evaluation of teaching can proceed. After data have been gathered, the task becomes combining it into a usable form.

The examples below use a common 1 to 4 scale, with 4 as the highest rating and 1 as the lowest. All forms, including questionnaires, interview schedules, and any other measurement tools used to collect student ratings, peer ratings, and department head ratings report results on that scale. The same would be true if the 5-point scale or another measurement had been selected. Whichever scale is used, it must be consistent throughout the evaluation system.

Having determined the information to be provided by each source and specified the weights assigned to that information, it is now possible to compute an overall rating that reflects the collective values of the faculty. Each source provides information on teaching performance elements as previously determined by the faculty. The information from each source concerning each component of each teaching role is weighted in ways that reflect the consensus value system developed in collaboration with the faculty.

In other words, the overall rating or evaluation is based on the principle of controlled subjectivity (discussed in Chapter 4). Table 4 shows how data gathered by the various tools used to measure the performance elements of teaching for one faculty member might be assembled into an overall evaluation of teaching.

TABLE 5.6 Weighting Measurement Data to Produce an Evaluation of Teaching

Minimum 20%		**TEACHING**		**Maximum 60%**	
Performance Component	**Sources of Measurement Data**				
	Students (25%)	**Peers (45%)**	**Dept. Chair/ Supervisor (20%)**	**Self (10%)**	
Content expertise		4		4	
Instructional design	3	4		4	
Instructional delivery	4	3		4	
Instructional assessment	2	3	4	3	
Course management			2	3	
AVERAGE	3.0	3.5	3.0	3.6	**Weighted Sum**
WEIGHTED AVERAGE	**0.75**	**1.575**	**0.6**	**0.36**	**3.3**

The weighted sum shown in the right-hand column is the final evaluation of teaching for the instructor in this case. Ratings from each source of various teaching performance elements were averaged, and those averages were weighted in accordance with the values determined during the development of the evaluation system. Finally, the weighted averages were added together to produce the final evaluation. Using the principle of *controlled subjectivity* ensures an approximation of objectivity (i.e., consistency of conclusions based on the same data). Controlled subjectivity ensures the weights assigned to subjective values for each source of information are controlled in that they are consistent for each individual. The weighted sum of 3.3 in Table 5.6 indicates a favorable teaching evaluation.[1]

[1] For an in-depth discussion of this method of computing an evaluation of teaching, or an evaluation of faculty performance in any other role, see *Developing a Comprehensive Faculty Evaluation System*, 3rd ed., by R.A. Arreola (Jossey-Bass, 2007).

Finally, as the heading of Table 5.6 indicates, the value range of the teaching role in the overall evaluation ranges from 20 percent to 60 percent. This range was determined by the faculty in accordance with the faculty role model (see Table 4.1) developed for the larger institution. Thus, in the overall evaluation of an instructor for decisions involving promotion, tenure, or merit pay, the evaluative outcome for teaching may be given a different weight than the weights assigned to *scholarly and creative activities* and *service*—the other components of an overall faculty role.

For example, suppose the faculty member whose data are shown in Table 5.6 had a professional assignment that included not only *teaching* but also various forms of *scholarship* and *service*. Suppose, then, that the roles relative to his or her specific professional responsibilities were represented as follows:

- teaching (45 percent)
- scholarly/creative activities (45 percent)
- service (10 percent)

If all faculty evaluations for the institution used the same 4-point scale, the evaluation of the scholarly/creative activities and service component would result in values similar to the value for teaching. For example, in addition to the teaching evaluation of 3.3 shown in Table 5.6, suppose the faculty member received an evaluation of 3.6 for scholarly/creative activities and 2.7 for service. The overall evaluation could then be computed as shown in Table 5.7.

TABLE 5.7 Computation of an Overall Faculty Evaluation

Role	Weight	Evaluation	Weighted Evaluation
Teaching	45%	3.3	1.478
Scholarly/creative activities	45%	3.6	1.620
Service	10%	2.7	0.270
		Weighted sum (overall evaluation)	**3.4**

The development of the metric for the evaluation of teaching as shown in tables 4.2, 5.1, and 5.6, as well as the metric for the overall evaluation shown in Table 5.7, provide a consistent mechanism for using faculty evaluation data in promotion and tenure decisions, as well for determining the allocation of merit-pay dollars:

- The standards for awarding a promotion could be set in terms of a specific overall evaluation value for a certain number of years.

- The standards for qualifying for tenure could be set in those same terms plus the achievement of a specific evaluation value in any of the roles, including, of course, teaching. However, in the awarding of tenure, faculty performance is usually only one of a number of factors taken into account.

- The standards for determining merit pay could be set in terms of achieving a specified minimum value in both an overall evaluation and specific role evaluations.

LINKING EVALUATION AND PROFESSIONAL ENRICHMENT

Faculty evaluation and professional enrichment are two sides of the same coin. Ideally, faculty evaluation programs and professional enrichment programs should work hand in hand. For example, if a particular aspect of faculty performance is being evaluated, faculty should have access to resources or opportunities to gain or improve the skills necessary for that aspect of performance.

For maximal self-improvement effect, *faculty-evaluation systems should be linked to, but operationally separate from, professional enrichment programs*. As a rule of thumb, if a specific aspect of faculty performance is to be evaluated, resources should be available to enable faculty members to gain expertise and proficiency in the skills required for that performance component—especially if that performance area is outside the area of engineering. Professional enrichment programs in educational psychology, instructional technology, conflict management, public speaking, and organizational management, for example, may assist faculty in achieving excellence in the full range of their professional performance.

The experience of the committee members indicates that no matter how well faculty evaluation systems are designed, if they are implemented without reference to opportunities for professional enrichment, they are inevitably considered primarily punitive. In addition, professional enrichment programs that are implemented without reference to the information generated by faculty evaluations tend to have disappointing results, no matter how well the programs are designed and funded. This situation is neatly summarized by Theall's (2007) statement that, "Evaluation without development is punitive. Development without evaluation is guesswork."

The reason is simple, if not always obvious. Unless professional enrichment programs are linked to evaluation systems, they tend to attract primarily faculty who are *already* motivated to seek out resources and opportunities to improve their skills. In short, the "good" seek out ways to get better—which is the quality that tends to make them good in the first place. However, individuals who are not thus motivated, and who, accordingly, are probably in greatest need of professional enrichment opportunities, generally tend to be the last ones to seek them out. Leadership from deans and department chairs can create an atmosphere of continuous improvement regarding teaching effectiveness by engaging the faculty in ongoing discussions about teaching (and related activities) as a pursuit of excellence.

When the elements of faculty evaluations are carefully coordinated with a professional enrichment program, the institution is more likely to obtain a valuable benefit from both. Thus, if an instructor's skill in assessing student learning is going to be evaluated, the institution should provide resources and training opportunities for him or her to become proficient in that skill. If a faculty member's ability to deliver a well organized, exciting lecture is going to be evaluated, resources should be available for him or her to become proficient in the requisite public speaking and presentation skills.

We must keep in mind that most instructors have had little or no formal training in the complex, sophisticated skills involved in designing and delivering instruction or assessing student learning outcomes. Most tend to teach the way they were taught and test the way they were tested. Thus, if faculty performance is evaluated, especially performance in teaching, the institution should provide resources for educators to develop, support and enhance their teaching performance.

In summary, a successful faculty evaluation system must provide (1) meaningful feedback to guide professional growth and enrichment *and* (2) evaluative information on which to base personnel decisions. The key to a system that serves both of these purposes is in the policies that determine the distribution of the information gathered for evaluations.

As a general principle, detailed information from questionnaires or other evaluation tools should be provided exclusively to the faculty member being evaluated as a guide to professional enrichment and growth in certain areas. However, aggregate data that summarize and reflect the overall *pattern of performance of an individual over time* should be used for personnel decisions, such as promotion, tenure, continuation, and merit raises.

It is important that everyone, both faculty and administrators, understand that evaluation data will be used *both* to provide faculty with diagnostic information to encourage their professional growth *and* to provide administrators with information that will be used in making personnel decisions (promotion, tenure, pay raises, etc.) An institution may emphasize one use over another, but it would be a mistake to pretend that faculty evaluation data will only be used for professional enrichment purposes. And, even if the primary intent is to use evaluations for professional enrichment, they should be designed so they can also be used for personnel decisions.

6

Conclusions and Recommendations

Institutions that have developed programs and support structures to enable faculty in engineering and other disciplines to improve their teaching skills have found that instructional enhancement programs on campuses often have high enrollments, and are sometimes even over-subscribed. However, because these activities are optional and because of limited institutional resources/capacity, participation among engineering faculty is relatively low and uneven.

Faculty in all disciplines must continually prioritize their time to reflect their most urgent faculty obligations, but also to prioritize their efforts in ways that will improve the prospects of career advancement. The current perception is that research is the most important aspect in faculty promotion and tenure decisions, because research contribution drives the market for faculty hiring and advancement. In addition, methods for assessing research accomplishments are well established—even though imperfect, whereas metrics for assessing teaching, learning, and instructional effectiveness are not as well defined or well established.

The development of a thoughtfully designed and agreed-upon method of evaluating teaching effectiveness—based on research on effective teaching and learning (NRC, 1999)—would provide administrators and faculty members with the wherewithal to use quantitative metrics in promotion and tenure decisions. Such metrics would also provide individual faculty members with an incentive to invest time and effort in developing their instructional skills, because they would be favorably reflected in advancement decisions.

Developing metrics for evaluating instructional effectiveness should be undertaken with the understanding that all faculty and the administration will have significant input into the design of the evaluation system, as well as feedback from the results. The assumptions, principles, and expected outcomes of the evaluation method should be explicit (and repeated frequently) to those who will be subject to evaluations, as well as those who will participate in administering the evaluations.

The model in which the department chair serves as both the first-line administrative evaluator and primary faculty development officer is not tenable and generally not recommended (Arreola, 1997). For that reason, the gathering process and use of the assembled information for administrative (tenure and promotion) evaluations should be decoupled from information gathered for use in professional development (although it is recognized that the types of information for each purpose are likely to have significant overlap) in order to foster an atmosphere where faculty can engage in professional development activities free of concerns that identifying weaknesses could reflect negatively during administrative evaluations. The development of agreed-upon metrics will also provide accrediting agencies with an added means of assessing instruction.

Information for evaluations of teaching should not be limited to student ratings, which address only the aspects of teaching students can observe. Other methods of evaluation, such as peer reviews of the quality of instructional design and content (along with self-evaluations and evaluations by department heads of those same items) can lead to a fuller understanding and more useful assessment of instructional effectiveness. Specific metrics and procedures are outlined in Chapters 4 and 5 of this report.

The following recommendations provide guidelines and specific actions to assist institutions and other stakeholders in developing and deploying metrics for instructional evaluations that will be widely accepted and relevant to engineering faculty.

Recommendations for Institutional Action

Institutions, engineering deans and department heads should:

1. Use multidimensional metrics that draw upon different constituencies to evaluate the content, organization, and delivery of course material and the assessment of student learning.
2. Take the lead in gaining widespread acceptance of metrics for evaluating scholarly instruction in engineering. Their links to faculty and institutional administrators give them the authority to engage in meaningful dialogue in the college of engineering and throughout the larger institution.
3. Seek to ensure appropriate quantities of evaluators who have the knowledge, skills, and experience to provide rigorous, meaningful assessments of instructional effectiveness (in much the same way that those institutions seek to ensure the development of the skills and knowledge required for excellent disciplinary research).
4. Seek out and take advantage of external resources, such as associations, societies, and/or programs focused on teaching excellence (e.g., Carnegie Academy for the Scholarship of Teaching and Learning, Higher Education Academy (U.K.), and Professional and Organizational Development Network), as well as on campus teaching and learning resource centers and organizations focused on engineering education (e.g., International Society for Engineering Education [IGIP] and the Foundation Engineering Education Coalition's web site on Active/Cooperative Learning: Best Practices in Engineering Education http://clte.asu.edu/active/main.htm).

Recommendations for External Stakeholders

Leaders of the engineering profession (including the National Academy of Engineering, the American Society for Engineering Education, ABET, Inc. The American Association of Engineering Societies, the Engineering Deans' Council, and the various engineering disciplinary societies) should:

1. Continue to promote programs and provide support for individuals and institutions pursuing efforts to accelerate the development and implementation of metrics for evaluating instructional effectiveness.
2. Seek to create and nurture models of metrics for evaluating instructional effectiveness. Each institution, of course, will have particular needs and demands; however, nationally known examples of well informed, well supported, and carefully developed instructional evaluation programs will benefit the entire field.

References

ABET. (2008). 2008-09 Criteria for evaluating engineering programs. Available online at http://www.abet.org/Linked%20Documents-UPDATE/Criteria%20and%20PP/E001%20 08-09%20EAC%20Criteria%2011-30-07.pdf.

Aleamoni, L. M. (1981). Student ratings of instruction. In J. Millman (Ed.) *Handbook of teacher evaluation* (pp. 110-145). Beverly Hills, CA: Sage Publications.

Arreola, R. A. (1997). On the tightrope. *The Department Chair,* 7 (4). Boston, MA: Anker Publishing Co.

Arreola, R. A. (2007). *Developing a comprehensive faculty evaluation system* (3rd ed.). San Francisco, CA: Jossey-Bass.

Arreola, R. A., Theall, M., & Aleamoni, L. M. (2003). Beyond scholarship: Recognizing the multiple roles of the professoriate. Paper presented at the 2003 American Educational Research Association Convention, Chicago, IL. Available online at http://www.cedanet.com/meta/Beyond%20Scholarship.pdf.

Atman, C. J. (2007, June 8). *Engineering education research: Some history and examples from the U.S.* Opening address to the Danish Centre for Engineering Education Research and Development, Copenhagen, Denmark. Available online at http://www.engr.washington.edu/caee/Atman%202007.06.08%20Denmark-final.pdf.

Braskamp, L. A. & Ory, J. C. (1994). *Assessing faculty work.* San Francisco, CA: Jossey-Bass.

Brint, S. (2008). The Spellings Commission and the case for professionalizing college teaching. *ACADEME Online*. Available at http://www.aaup.org/AAUP/pubsres/academe/2008/MJ/Feat/brin.htm.

England, J. (1996). How evaluations of teaching are used in personnel decisions. American Council of Learned Societies, Occasional Paper N. 33. Available at http://archives.acls.org/op/33_Professonal_Evaluation_of_Teaching.htm.

Felder, R. (1993). Reaching the second tier: Learning and teaching styles in college science education. *Journal of College Science Teaching,* 23(5), 286-290.

Lencioni, P. M. (2002). *The five dysfunctions of a team: A leadership fable.* San Francisco, CA: Jossey-Bass.

Michell, J. (1997). Quantitative science and the definition of measurement in psychology. *British Journal of Psychology,* 88 (3), 355-383.

National Academy of Engineering. (2005). *Educating the engineer of 2020: Adapting engineering education to the new century.* Washington, DC: The National Academies Press.

National Research Council (1999). *How people learn: Brain, mind, experience and school.* Washington, DC: National Academy Press.

National Research Council. (2003). *Evaluating and improving undergraduate teaching in science, technology, engineering, and mathematics.* Washington, DC: The National Academies Press.

National Science Foundation. (2002). Science and engineering indicators 2002. Available at http://www.nsf.gov/statistics/seind02/c3/c3s2.htm.

Paese, P. W., Lind, E. A., & Kanfer. (1988). Procedural fairness and work group responses to performance evaluation systems. *Social Justice Research 2(3)*, 193-205.

Theall, M. (2007, June 11). *Basic issues in good evaluation practice.* Invited presentation at the Academic Impressions Institute's Using Faculty Evaluation Data for Decision Making, Boston, MA.

United States Department of Education. (2006). A test of leadership: Charting the future of U.S. higher education. Washington, DC: USDEd.

University of Texas. Preparing for peer observation: A handbook. Retrieved December 1, 2008, from http://www.utexas.edu/academic/cte/PeerObserve.html#AppendixList.

Van Note, C. H., & Szabo, B. (1996). Who uses faculty development services? In *To Improve the Academy*, vol. 16. Stillwater, OK: New Forums Press, Inc.

Van Prooijen, J-W. (2007). Fairness judgments: Genuine morality or disguised egocentrism? In Mind Articles, Special Issue 5. Available online at http://www.in-mind.org/special-issue/fairness-judgments-genuine-morality-or-disguised-egocent.html.

Appendixes

Appendix A

Workshop Agenda and Attendees

Workshop on Metrics of Instructional Scholarship

November 13, 2007
National Academy of Science Building
2100 C Street, NW, Washington, DC
Room 150

Meeting Agenda:

8:30 am	Welcome and Opening Comments by Study Committee Chair *C. Judson King – University of California Berkeley*
8:45 – 10:15	Presentations regarding strategies for developing candidate metrics, assessment or scoring, and evaluation agents. • *Larry Braskamp – Loyola University of Chicago* • *Lawrence Aleamoni – University of Arizona* • *Michael Theall – Youngstown State University* • *John Bardo – Western Carolina University*
10:15 – 10:30	Break
10:30 – 12:30 pm	Breakout Discussions – Breakout Groups will address the points raised in presentations and augment those ideas in order to propose methods to devise a metric/rubric. Community Acceptance would be dealt with by all groups of discussants.
12:30 – 1:30	Lunch – Continue breakout group discussions
1:30 – 2:30	Brief Presentations from each of the breakout groups
2:45 – 4:30	Plenary Discussion – Moving toward group consensus for Metric Development, Scoring, Assessment/Evaluating Body and Community Acceptance
4:30	Adjourn

Workshop Attendees

Lawrence Aleamoni, University of Arizona
Susan Ambrose, Carnegie Mellon University
Brownell Anderson, Association of American Medical Colleges
Raoul Arreola, University of Tennessee
John Bardo, Western Carolina University
Larry Braskamp, Loyola University of Chicago
Elizabeth Cady, National Academy of Engineering
Mark Fleury, National Academy of Engineering
Norman Fortenberry, National Academy of Engineering
Kamyar Haghighi, Purdue University
Susan Kemnitzer, The National Science Foundation
Julia Kregenow, National Academy of Sciences
Thomas Litzinger, The Pennsylvania State University
Jack Lohmann, Georgia Institute of Technology
James Melsa, American Society for Engineering Education
Lueny Morrell, Hewlett Packard Company
Wilfrid Nixon, American Society of Civil Engineers
George P. "Bud" Peterson, The University of Colorado
M.P. Ravindra, Infosys Technologies, Limited
Paul Savory, University of Nebraska
Allen Soyster, National Science Foundation
Richard Taber, National Academy of Engineering
Mike Theall, Youngstown State University
Elizabeth VanderPutten, National Science Foundation
Thomas Walker, Virginia Polytechnic Institute and State University
Robert Warrington, American Society of Mechanical Engineers
Karan Watson, Texas A&M University

Appendix B

Biographical Sketches of Committee Members

C. Judson King (chair), is director of the Center for Studies in Higher Education and professor emeritus of chemical engineering. At the Center, his research focuses on systemic and institutional concerns as well as issues specific to engineering and technical disciplines. His chemical engineering research has centered upon separation processes, including spray drying, freeze drying, solvent extraction, and adsorption. Since joining the University of California in 1963, King has served in a variety of academic and administrative posts on the UC Berkeley campus and the system level. Most recently, he was Provost and Senior Vice President – Academic Affairs of the University of California system (1995-2004), and before that system-wide Vice Provost for Research. At UC Berkeley, King served as Provost – Professional Schools and Colleges, dean of the College of Chemistry, and chair of the Department of Chemical Engineering. He is a professor of Chemical Engineering and has written over 240 research publications and a widely used text book, *Separation Processes*. Professor King is a member of the National Academy of Engineering and has received major awards from the American Institute of Chemical Engineers, the American Chemical Society, the American Society for Engineering Education and the Council for Chemical Research. He has been active with the California Council on Science and Technology.

Susan Ambrose is Associate Provost for Education, Director of the Eberly Center for Teaching Excellence, and Teaching Professor in the Department of History at Carnegie Mellon. She received her doctorate in American History (1986) from Carnegie Mellon and has been on the Eberly Center's staff since its inception. She has designed and conducted seminars and workshops for faculty and administrators throughout the United States and in India, Canada, Mexico, Taiwan, South Korea, Hong Kong and Chile. In 1998 and 2000 she was named a Visiting Scholar for the American Society of Engineering Education and the National Science Foundation, spending time with the engineering colleges at the University of Washington-Seattle, Rice University, and Tufts University. She was also awarded an American Council on Education fellowship for 1999-2000 and worked alongside the presidents of Connecticut College and the University of Rhode Island to learn more about leadership styles. She has received funding over the years from the National Science Foundation, the Alfred P. Sloan Foundation, the Fund for the Improvement of Postsecondary Education, the Lilly Endowment, the Carnegie Corporation of New York, and the Eden Hall Foundation to conduct research on women in engineering and science and create support programs for targeted groups such as first-year engineering faculty and women and minority faculty in science and engineering. She is co-author of *The Woman's Guide to Navigating the Ph.D. in Engineering and Science* (2001) with Barbara Lazarus and Lisa Ritter; *Journeys of Women in Engineering and Science: No Universal Constants* (1997) with Kristin Dunkle, Barbara Lazarus, Indira Nair and Deborah Harkus; *The New Professor's Handbook: A Guide to Teaching and Research in Engineering and Science* (1994) with Cliff Davidson; and numerous chapters and journal articles. She also teaches courses on immigration and ethnicity in the Department of History at Carnegie Mellon.

Raoul A. Arreola received his Ph.D. degree in Educational Psychology, specializing in measurement and evaluation, in 1969 from Arizona State University. Teacher, author, trainer and consultant to nearly 300 of colleges nationally and internationally, Dr. Arreola has published in the areas of distance education, academic leadership, and faculty evaluation and development. His best-selling book *Developing a Comprehensive Faculty Evaluation System* now in its second edition, is widely used in colleges and universities in designing faculty evaluation programs. Dr. Arreola is on the faculty of The University of Tennessee Health Science Center where he has also held several administrative positions including Chairman of the Department of Education and Director of Institutional Research, Assessment and Planning. In addition Dr. Arreola has served on the staff of the University of Tennessee Institute for Leadership Effectiveness as one of only 15 faculty selected from throughout the University's 5-campus system to serve as a facilitator in the leadership training of over 300 academic administrators. For the last 17 years Dr. Arreola has conducted national workshops on faculty evaluation for thousands of faculty and administrators from over 500 colleges throughout the world. He has been invited to make numerous keynote addresses on the topics of assessing faculty performance, evaluating and enhancing teaching, the use of technology in teaching, and identifying, measuring, and developing he skill set components of teaching excellence. For the last two years Dr. Arreola has been a featured presenter on faculty evaluation in Magna Publication's Audio Conference series. In 2004 Dr. Arreola received the *McKeachie Career Achievement Award* from the American Educational Research Association's Special Interest Group on Faculty Teaching, Evaluation, and Development. In addition, in 2005 his work on defining the professoriate as a 'meta-profession' and identifying the subordinate skill sets of faculty work was recognized by the American Educational Research Association which awarded him its prestigious *Relating Research to Practice Award.*

Karan L. Watson, Ph.D., P.E. is interim Vice President and Associate Provost for Diversity at Texas A&M University. From February 2002 through November 2008, she served as Dean of Faculties and Associate Provost since February 1, 2002. She joined the faculty of Texas A&M University in 1983 in the Electrical Engineering Department, where she is currently a Regents Professor. Dr. Watson is a registered professional engineer and has been named a fellow of the Institute of Electrical and Electronic Engineers (IEEE) and the American Society for Engineering Education (ASEE). She received the US President's Award for Mentoring Minorities and Women in Science and Technology, the American Association for the Advancement of Science (AAAS) mentoring award, the IEEE International Undergraduate Teaching Award, the TAMU Association of Former Students University Award for Student relationships, the TAMU Provost's Award for Diversity, the TAMU Women's Week Award for Administrators, the College of Engineering Crawford Teaching Award, and was named a TAMU Regents Professor. She has chaired the doctoral committees of 32 students and over 60 master degree students. In 2003-2004 she served as a Senior Fellow of the National Academy of Engineers' Center for the Advancement of Scholarship in Engineering Education, and since 1991 she has served as an accreditation evaluator and commissioner for engineering programs for ABET, both in the US and internationally. In November 2005 she was named as the Interim Vice President and Associate Provost of Diversity.